Zahra M.M.A. Sadiq

Computer Music

Electronica, Algorithms, Artificial Intelligence

Music is the language
of the universe's
invisible art

The contemporary musician

Compositions of the late twentieth and early twenty-first century which realize a theoretic conception of music represent contemporary sonic art. As a composer I create individual techniques and rules and I define my specific notion of music. My sonic and theoretic oeuvre is a contribution to the development and refinement of contemporary compositional conceptions. Musicians who master their craft are extraordinary, intelligent and creative and their knowledge is logical and mathematical as well as theoretical and compositional.

I am a pioneer, creating the conception for a new genre, and a specialist for historical and contemporary music. I acquire and develop knowledge in a field of musical science that is still unknown. Because the historical and the prospective systems are self-contained, the combination of both structures and methods is not possible. I have solved the contradictions by deconstructing the concrete forms and by forming abstract sets of functions that I have reclassified and refine. Based on this research I have constructed a new musical system.

Avant-garde

For me, to be a pioneer means to create a path by using it and to share it with the listeners all over the world who appreciate this music because the compositions are extraordinary and the sound is exceptional. As a composer who is an innovator of new forms, I reinterpret the harmonic, melodic and rhythmic structure and thereby create a new sound language. I introduce a form of composition that reconciles historical and contemporary music of Western, Arabian, African and Asian cultures on a theoretical level. This music is regarded as avant-garde and among the qualities that make it exceptional are the unique and experimental forms and the reinterpretation of compositional techniques.

The oeuvre of a musician who contributes to the development and refinement of compositional conceptions encapsulates the past and shapes the future. I define criteria for good music of the late twentieth and early twenty-first century. The compositional forms that I realize are multi-faceted and include historical techniques and structures that I integrate into the contemporary conception. The highly elaborated music of various ages and countries belongs

2

to systems which are to a large extent structurally self-contained. In every age the emergence of these forms of music is advanced by innovations that concern the technique of musical instruments. The invention of electronic instruments, the computer and its software enable the programming of algorithmic music, and hybrid forms combine direct programming procedures and automated processing. Another important influence is the compositional technique that I create by carrying out scientific research and by discovering mathematical and acoustic rules. Changes in the social context also contribute to modifications of the compositional process.

The theory of sound

The relationship between theory and reality is complex. The cosmos and geometry, music and music theory appear to be opposites. Cosmology and music are inseparably based on mathematical and logical principles. I compose contrapuntal music in the form of algorithms and assume that the calculation of frequencies is determined by cosmological relationships. If a musical theorist invents a new calculation method for scales and proportions, composers create their music by applying the new sound system.

Musicology describes characteristics and similarities of pieces of music from a historical perspective. The classification of composers and their music into epochs and genres describes the theoretical basis for compositional contributions to the history of music. As a musician who is a pioneer creating a new genre I consider and implement the entirety of the arts of calculating scales and proportions and developing and structuring new micro and macro forms in order to make a unique contribution to music theory. These compositions will have an influence on the work of musicologists, music theorists, composers and audio engineers of tomorrow.

The floating counterpoint

The genre floating counterpoint integrates an interdisciplinary conception that I apply in the context of creating eai music of the twenty-first century. The term eai music designates a form of music that combines electronic and computational algorithmic procedures including artificial intelligence. In order to create a concept for this specific form of music I redefine the numerical proportional relationships and develop a conception that integrates a compositional, an esthetic and a technical structure.

4

A large number of compositional techniques, forms and genres has been developed and continues to be created. As someone introducing a new genre, I receive no support when I decide to shape the form of music and I often face much criticism. At the same time, I am less comparable to other extraordinary composers and I am free to develop a new syntax, semantics and sound, and new forms of music. After it becomes established this music will shape compositions of the highest order with its exceptional quality.

An infinite number of words is not sufficient to define the music of a composer, an era or even a single piece of music. The language of music is formalized in the same way as the language of mathematics, it is only expressed by sounds. In spite of this, it is possible to describe certain similarities and characteristics of genres and sub-genres. The floating counterpoint genre designates a form and technique of composition that reconciles the historical and contemporary music. It is the reinterpretation of compositional techniques characterized by colorful harmonics, the extended tonality of the early moderns and contemporary music, by elaborate counterpoint and highly structured serial techniques. Music based on serial methods or counterpoint appears determined but the choice of sounds is connected with their integration into the entire conception of floating counterpoint music. I also create music based on mathematical patterns in the form of fractals or numerical sequences. The numerical values of the fractals are mapped to sonic parameters. As a composer, I choose the values for the iterative mathematical formulas in a way that determines proportionate and harmonic patterns. I associate it with the compositional, sonic and esthetic conception of the floating counterpoint.

Floating counterpoint integrates ancient oriental and Pythagorean concepts. This implies that the world's being is the harmonious compound of infinite and finite principles. This principle is rediscovered in the form of Torricelli's trumpet, a geometric figure possessing an infinite surface area but a finite volume.

The floating counterpoint genre defines a conception that structures the constellation of the musical elements. It is the technique of an innovative approach to the meta-level with a complex structure. It designates the shifting functionality of elements at the macro and micro levels and their patterns. The macro level pattern refers to the system and interplay of scales, medieval modi, hexachords, counterpoint, diastematic structures and the rhythm, which all affect each other. I create a musical system that presupposes a pool of tones connected with a modulation scheme, a scale and a system that regulates the harmonies. This is also valid for scales which are not tuned in an even-tempered way. The micro level pattern relates to a short section of music and it is identical with the soggetto, the source from which other forms are derived. The macro level pattern and the micro level pattern are interdependent and develop different constellations. I modify the interplay of the musical dimensions by transforming the macro level pattern. These changes affect the micro level pattern because the development of the soggetto, which is an important musical idea, is based on the relationship of harmony, melody and rhythm with each other. Modifications of the macro level influence the micro level and the microform contributes to the macro-form. Both are part of the entire compositional conception and its rules. It depends on the composer, his knowledge and ability to weight and shape the dimensions. The musical fundamentals of floating counterpoint music hover and I determine their significance by redefining their function systematically in the compositional contrapuntal conception.

7

Serial techniques of the floating counterpoint principle are not limited to sequences of sounds which are not modifiable. I define serialism and determine rules for the use of scales and parts of them and their combination that is a horizontal procedure. The scale is the center of the composition and its processing determines the micro and macro patterns of the music. My compositions avoid connecting dissonances synchronously and successively. I regulate vertical harmony and polyphony by integrating the rules of the counterpoint, the pool of tones and the modulation scheme. The modal structure influences the harmonic and serial conceptions. I prefer the techniques of the medieval sound system, unanimous and polyphonic counterpoint, serial and diastematic techniques and the combination of two or more keys. A polyphonic and contrapuntal texture presupposes that the parts of the score possess a soggetto, harmonics and rhythm that are independent and to some extent self-contained. I supplement the cantus firmus with parts that act as a counterweight and integrate the important chord progressions of this polyphonic music into the movement of the parts. I realize rhythmic structures immanently.

I use an instrument to produce oscillations whose frequencies divide the octave infinitely many times. When I create a sound starting with the smallest acoustic and physical component by applying generators, this procedure is part of preparing the composition. As a composer I combine sine waves, waveforms with an abundance of overtones and complete white noise. When I use random frequencies to compose a piece of music, the structure of the music is based on probabilities. I integrate this conception into a determined structure of the composition. It has the function of an improvisation and extends the structure of the music in a way that is coherent rather than random. I integrate the natural sciences into the conception of the composition and the creation of the tones forms part of the music of the spheres.

When I compose bitonal or even tritonal pieces of music, it is possible to explain the context of the harmonies according to functional harmonics. I elucidate each of the harmonic keys by emphasizing certain cadences. Today there are still limitless possibilities for the composer to create music by integrating harmonic rules. It is still possible to use creative methods for modulating or combining different keys.

In the twentieth and the twenty-first century, the tonal implication gave way to patterns, counterpoint and formal transformations. The floating counterpoint concept structures the syntax and semantics of a language called music. Repetitions organize the musical material but they do not form part of it. When I compose a piece of music and apply the rules of harmonics, I use the art of modulation to cause movement. If I treat cadences like caesuras, the soggetto and its harmonics become rhythmically more independent. Regular forms and proportions and the constant division of time structure the music. Symmetries and the proportional relationship between the sounds contribute to the beauty of the music. Floating counterpoint is the reduction of music to its essence. It is very detailed and focuses on the essential at the same time. It is a concise form of music, possessing an austere form that is refined by a strict counterpoint and leaving only limited possibilities for frequencies not integrated into the serial system. Regularity and deviation, movement and deceleration, densification and tension release interact.

The musical process of my compositions is not completely determined; I create musical patterns or sogetti which introduce new harmonies, rhythms or

1 0

melodies into the piece of music. Some of these are weakly connected with the existing material and the structure of the composition. If parts of floating counterpoint music are composed around the centre of one or more musical ideas, the relationship between the subjects is not exclusive but integrative. The macro-form of the piece of music depends on the microforms and when it is combined with the nuba or the sonata, it is modified, abstracted and assimilated with the conception of the floating counterpoint that is structured by its internal logic and contrapuntal patterns. I create innovative forms that are regular and symmetrical, although exceptions occur, and I transform, compress and extend the musical patterns. The piece of music is divided into one or multiple parts and it is especially tripartite with extensions that contribute to the structure and it can be recurrent.

Because floating counterpoint music is conceptual and procedural, it is open for improvisation and development. This form of spontaneous invention is only possible within the framework of the existing conception and patterns. It is a defined form of improvisation that follows strict rules. It is possible to create a piece of music belonging to this genre which is exclusively composed and performed by a program. This form of computer music is admirable because the output is not completely controlled, despite being trained to produce a coherent and variable structure. It is interesting to combine this processing method with the musician's creativity, experience and the social context of the musical performance. I integrate improvisation and music that is mechanical, electric-analogue and electronic-digital.

The processual method is part of compositional technique and it is not explicit. It is part of the structure of the musical material and it forms a language that I use to express ideas on an abstract and indirect level. Single sounds or phrases are related to other sounds and the entire composition on macro and micro levels. Although nearly every sound of the austere form is related to other sounds, the composition is at no point determined. The composer introduces new material, extends the structure and modifies the patterns. I realize sounds according to their statistical relevance in the framework of the predefined rules.

Independently of the application of indirect or direct processing methods, the composition follows the rules of floating counterpoint. Indirect programming and the use of artificial intelligence contributes to the creation of algorithmic music. I combine methods of direct and indirect programming of the composition and sound editing methods. If the level of the procedure is sophisticated, it supports the quality of the entire composition.

Innovation

The recognition of sounds and pieces of music and their reminiscence contributes to the orientation of the listener in the context of experimental music. Innovation is the art of composing music which appears familiar while creating the unknown. Scales with the most irregular shape seem to be recognizable if the composition regards their proportional particularities while diminished or augmented intervals add alterations. Unusual forms appear conversant if the development of the musical elements, the patterns, soggetti and derivations is logical.

If the computer develops a form of music that I did not imagine during the process of programming, I analyze the result and integrate it into my existing sonic conception. If the program applies artificial intelligence and creates a piece of music which transcends the existing framework of the composition, I extend the structure and assimilate both ideas. The theory of the floating counterpoint, which is flexible and abstract, is suitable for this procedure.

Science and art

The universe and its scientific conception is the source of music and its theory. Mathematical and logical properties mediate between art and science. In the twenty-first century, artificial intelligence, interdisciplinary research and creative thinking are more important than in the centuries before. Form and content, technique and statement interplay. It is not important to avoid errors but to minimize and to master their influence. Complex thinking combines artistic and scientific reasoning, analytic and creative processes.

Arrangement and composition

Because the sound and structure of music correlate, I compose while imagining particular sounds. I apply algorithms to create sounds that result from algorithms and use physical and mathematical models that are obtained by analyzing the parameters of an instrument. The physical modeling of virtual instruments integrates concepts of artificial intelligence. When I program an abstract conception that creates the sound and structure of the music, the music of the spheres constitutes the scientific principle that determines the frequencies. This concept involves the analysis of the cosmological proportions and their transfer to the musical realm. I integrate deviations from these frequencies that function as alterations into the composition.

The arrangement participates in the compositional process. I shape my rhythm sections in an unobtrusive and even ghostly form. If a percussionist participates in the performance, he stays in the background compared to his fellow musicians, who perform on melodic and harmonic instruments. Rhythm results from of the interaction of all instruments unless there is a solo. I blend the bass, the rhythm section and the harmonic and melodic process. The soggetti take part in the harmonic course and become two-dimensional, which is emphasized by imitational figures. I experiment with the self-contained function of compositional parts that I structure according to the contrapuntal rules. Because the concept of this procedure is flexible, I arrange parts that perform the sogetto and that are weighted similarly without changing their function at the macro level.

The musical measure consists of a certain quantity of equally weighted beats. Because of their uniformity, the beginning and the end of the phrase and its repetition is not perceived without contextual information. In music other than freestyle, the rhythmical emphasis that is supported by the soggetti and harmony transforms the physical time into measures and musical patterns. I integrate concrete sounds into my compositions to broaden the scope of the rhythm section and I transform them into musical elements. I also apply them as a physical signal that triggers the synthesizing process of sound shaping. Since my music is abstract, it does not reflect realities.

The arrangement contributes to the inimitable sound of floating counterpoint music that is described as reflective, sparkling, pristine, delicate and spherical. It is cool, reduced and suspenseful, and realizes opposites and diversities. Floating counterpoint music is a form of chamber music and it is concise and complex. This does not depend on the way I realize the composition with mechanical and electrical instruments, with the synthesizer or in a hybrid form. I limit the loudness of the digital mix and master in order to emphasize the details and the dynamic differences between the independent parts of the score. If the music is performed with conventional or digital instruments or in a hybrid form, I arranged it with a solo instrument, multiple instruments, a chamber orchestra or a grand orchestra. If the arrangement has the form of chamber music, the composition has a concise structure and it focuses on details and on the single parts.

Harmony

The theory of musical texture and harmony is defined by rules that attribute properties to sounds. The perception of its harmonic message is determined by the context and multi-layered references. While the structure constitutes ambiguities for the program and for the musician, it characterizes the sophistication and complexity of the harmonic language and the implicit multifaceted structure of the contrapuntal composition for the musician.

Standards

Folk music and artificial music are connected and often it is hard to tell the true nature of a soggetto. As a composer I study the standards but I do not process them into my original compositions. From my point of view, the history of music is a pool of different eras with specific coherent musical systems in different periods and countries from which I can abstract and transform to create a new structure and conception.

World music

The cultural diversity and individuality of the compositions that characterize world music presuppose an exceptional degree of musical proficiency. World music is divided into genres and sub-genres and it blends different styles, instruments and a variety of influences. Musicians combine influences from classical music, blues and folk music from different continents and cultures, integrating different ways of composing and experimenting with harmony, melody and rhythm. The forms and compositional techniques they develop are unique and innovative. They combine new and traditional methods of composing, and conventional and electronic instruments meet in a harmonious manner. The experience of world music artists extends beyond mechanical, electric and computer music, transcends ancient and contemporary music and music of all continents.

Sound worlds

Music is the sound of periodicity and proportion in time and space. It is invisible and the particular characters of expression are experienced by sensory and cognitive perception. According to Pythagoras and his predecessors from the ancient orient, the whole cosmos is number and harmony. Music enables beings on earth to appreciate the sound of the universe and to listen to the cosmic music of the spheres. Beings anywhere in the universe appreciate proportions emerging from the mathematical regularities of space.

Planets radiate electromagnetic energy and waves and their music consists of regular movement that coincides with musical frequencies. My compositions emphasize these cosmic sounds and proportions and place them in the centre of the music or parts of it. The listener travels through space and listens to a composition that the cosmic electronic ensemble performs. If I integrate cosmic octaves and harmonies, I emphasize certain perennial periods of the cosmic constellation. Floating counterpoint music is music of the universe.

Silence

Silence is the sound of the musician. Music is the sound of the cosmic sphere.

The art of listening

Floating counterpoint music is appropriate for listening purposes. The rhythm is part of the pattern that structures the music and is therefore not a dance rhythm. The audience listens silently while the music is performed and the room responds.

As a musician who composes and produces music by using score editors, digital audio workstations and artificial intelligence-based programs, I train my capacity to listen. I learn to imagine melodies, sounds, polyphonic and harmonic or contrapuntal ideas. When I program algorithms to create a composition that applies procedures of artificial intelligence, I formulate sonic ideals and rules, train neural nets and assess their results. The compositional process shifts to the preparatory phase of the composition and the intermediate and final stage of evaluation.

Rooms

Sound needs a perfect room. Reverb simulates the nature of a room. It is possible to perform floating counterpoint music electronically, unplugged or in a blended form. The appropriate musical location for performing electronic music is the internet, a studio, a room or a concert auditorium. The description of the qualities of the ideal room that is most suitable for a musical performance is part of the creative and scientific process and is supported by electronic devices and computer programs.

Ensemble

As a studio musician, I compose music solely in my room. I make decisions, experiment with tonal systems and sounds. In order to compose a passage by activating the musical flow, I improvise with an ensemble or alone, developing a musical idea and creating a digital version of it.

In front of my computer I am free to compose passages that no virtuoso could perform. I create music whose performance demands an exceptional playing capacity and I use techniques that a mechanical and electrical and analogue instrument cannot imitate. An ensemble is nonetheless able to perform the pieces unplugged or in a blended form. In spite of the differences, it is possible to perform both versions in a way that emphasizes the sound of the specific instrument creatively.

Time

Timeless music is appreciated for centuries or even millennia. Even if the language of music changes, the oeuvre of the composer is acknowledged. Music of elapsed times is still performed and transformed. Compositions with an elaborated technical standard that express a deep understanding of the musical flow are timeless. The maintenance of contemporary electronic music depends on the technical devices it is stored and distributed with. The selection of the oeuvres which will be heard in the future depends on the tradition of notes, recordings, videos and data files. As far as complex compositional techniques are the result of an elaborated form of scientific ability and research, electronic music is a minor genre.

Time is precious for the musician who improvises on stage in real time. As a studio musician, I take the time that I need to compose, produce and program. Since music is closely connected with real time, improvising contributes to the flow of music even if it is performed during the process of creating and producing music in the studio. Musical time is dynamic and interdependently of real time, I define my compositional conception of time. Syncopated rhythms or the retarding or accelerated performance of parts of the score extend the temporal notion.

One of my compositions sounds like the movement of a moment that reflects only one specific sonic scene. Details only change inside the immobile framework of the sound section. Usually a piece of music which is based on a linear conception of time introduces soggetti, harmonies and patterns that the listener gets acquainted with in the course of the composition. These forms flourish, decay or change their levels because they develop in a temporal and harmonic sense. In polyphonic music the synchronic and diachronic development of the musical elements occurs simultaneously. I develop structures of floating counterpoint music that are abstract and individual and I realize a conception of time that is multi-layered. If I were to write a sonata form complying with the principles of floating counterpoint, it would be based on an abstract foundation rather than on a narrative style.

From a musical point of view, the pitch and length of a tone are not interrelated except for compositional reasons. The composer creates the rhythm, which is an artistic and relative measure that refers to time and weight. Proportions consist of tones that receive their musical meaning in relation to each other. The parameters of wavelength, frequency, velocity and amplitude constitute the physical framework that the musical realm transcends. These physical concepts are interrelated with sound concepts or even substitute them. Using algorithms to create music is connected with the flow of music, and the procedure is translated into its specific language. Although music is correlated with the sciences of physics and mathematics, it is a unique realm. The combination of the pitch and duration of a sound can be an effect of impulse generators and pitch shifting. The modification of the distances between the impulses of the impulse generator results in a change of pitch and the modification of the sound duration results in a change of the tone color. As a composer, I relate pitch and rhythm by programming an algorithm whose sonic result realizes the rules of music theory. I connect the pitch of a tone with the meter by using a pendulum with the length of a string belonging to an instrument and I measure its oscillation per minute. I calculate the frequency in hertz and the period. The resulting value constitutes the meter, which is also based on astronomical relations. Using this method, I advance and unify my music with the atmospherics, the oscillations of the universe, the wavelength of light and the resonance peak of the DNA. This procedure presupposes that the resulting rhythm is not only mathematical but also musical, and that it conforms to the floating counterpoint conception.

The art of producing

The art of producing electronic music with a computer is part of the performance and the musical interpretation that emphasizes the specific nature of the computer and the instruments. During the process of producing, I blend improvisatory passages and sections that are composed in detail. If I use a playback that I composed or generated by applying procedures of artificial intelligence to a part or to the entire composition, I enable the musician to align his performance towards the recording and in doing so to create a contrast to passages without playback.

Sophisticated musicians of many ages were proficient in composing and improvising. Programming and playing the classical guitar, the electric guitar or a guitar-to-MIDI-system, I mediate between improvisation and composition. When a computer program performs the improvisation, this form of reduction to the technical level is artistic if I integrate the automated part of the composition into my entire musical conception. In the process of creating music with algorithms of artificial intelligence, I define a conception of music that corresponds to the technique. Proportions and statistical values for the occurrence of single sounds or groups of sounds shape the structure of references. I refine and transform the technique and the compositional conception, which are interdependent.

When an ensemble performs this music, it interprets the pieces in different ways. On stage and in the studio it interprets the composition with electric or electronic instruments. The musicians even use mechanical instruments and create an unusual sound if the original production is purely electronic and digital. The performer shapes the sound of conventional mechanical and electrical instruments and lets them imitate electronic instruments.

Composing

Musicologists formulate statements about the past that do not predict the future. They abstract principles which they have detected by analyzing the oeuvre of different composers, cultures and ages. As a contemporary composer whose oeuvre will be examined by the music theory of tomorrow, I study tradition, create music and compositional techniques, develop instruments and forms, and experiment. The research of our schools describes the music in an asynchronous way. I integrate the esthetics, the historical and philosophical classification of music on the one hand and the compositional technique, the technology of music, its semantics and syntax on the other. Music is based on numbers and philosophy simultaneously.

I have studied the rules of music and its theory to such an extent that I am able to transform them and to develop my own research. The medieval sound system, new forms of tuning, or the creation of new instruments or technologies influence my work as a composer and music theorist. I reinterpret and refine the ambiguous structure and the interaction of musical elements. In the course of history, composers have created a vast variety of musical pieces with a small number of sounds. The essentials of music consist of time, pitch and proportion, or in terms of the musical language of rhythm, melody and harmony. Composing is the art of creating a sonic structure that is scientifically evaluable.

My music, which is tonal in a harmonic sense, establishes a relationship between consonances and dissonances and it includes medieval techniques. The proportion is the main element of the structure and I integrate it into variable musical conceptions. This is especially important for a composition that joins scales from the musical systems of different musical cultures. As a composer, I transform these structures before I integrate micro-intervals and formulate rules for their inclusion. The procedures of programming, composing and producing electronic music are interrelated. The arrangement, the mix and the code depend on the structure and conception of the composition.

Uncertainty and errors

Music is the energy of inspiration. It is the flow that transforms proportions and the compositional ability of a musician into a work of art. The performers of an ensemble cooperate with each other. As a composer of electronic music I spend much time coding computer software. I act jointly with the programs and accept their peculiarities. The collaboration involves a two-sided interplay between me as a composer and the computer. If the program is based on artificial intelligence, it does not align to my musical conception. It learns to evaluate the input and makes decisions. If I compose an extraordinary piece of music directly, the program is flexible and accepts my suggestions. The computer supports the musicians during the composition and the performance.

As a composer of contemporary music, I create structures that reflect my conception of sound. This procedure transforms rules that were developed in the course of musical history and applies feed-forward methods that innovate conceptual patterns. I establish paragons and I also choose an algorithm and a meta structure which changes and extends these structures dynamically.

Audio engineering is an artistic science. Like the art of composing, it creates rules that prove to be reasonable and that ensure sonic quality. In the grey areas which are not strictly defined and which are ambiguous for the computer, I enhance and transform these rules to generate a particular sound. These zones are the point of departure for creative development.

Conceptualization is a complex procedure and logic is a mode of thought. Errors are an inevitable part of innovation, artificial intelligence and quantum computing. In intelligent music, the error participates in the procedure of composing. By integrating error reduction methods and error transformation methods, I enhance the scope of the conception. Systems that function in spite of an error can be improved, and their ambiguities are a principle of complex scientific and artistic thinking. In comparison to deduction, inductive reasoning copes with uncertainty. Fuzzy logic, the calculation of probabilities and amplitudes of quantum algorithms are complex quantitative procedures that I do not completely control.

I determine the macro-form and microform of music by designing algorithms and define statistical values for the occurrence of a particular structure or sound. Tonal references are ambiguous and complex in the framework of the existing rules of music theory and of principles that I create as a composer. I develop the alteration of an input into an expected sound by applying multiple

3 O

strategies. I integrate methods that reduce errors into the process of creating a piece of music while performing classification procedures used by neural nets or quantum neural nets in order to represent the gradual process of approximating the composition with an ideal.

Electronica

The term electronica designates music that is produced and performed with electronic instruments. The art of composing is connected with the medium. If I abandon my guitar to work with my computer I want to experience the procedure of coding and the sound of a synthesizer to realize the imagination of a certain sonic model. I integrate and transform what I have learned about conventional music and create music that the computer performs. By composing and performing with hardware and software, I aim at developing and diversifying the specific compositional possibilities, the sounds and playing technique of this media. If the music that I have composed and arranged for a computer is performed by a mechanical or electrical and analogue instrument, the playing technique imitates the electronic performance as far as the instrument allows it.

Artificial intelligence, quantum computing, exoskeletons and computer control by brainwaves influence the integration of computers into the process of composing and producing. Electronica is the form of music which reflects this development and contributes to its refinement. Music is a quantitative science and art, and the integration of mathematical and physical procedures into the compositional process is a complex process. If contrapuntal or serial structures shape the music, it appears determined, but the choice of sounds is connected with blending it with the entire technical and compositional conception of floating counterpoint music.

Synthesizer

The creation of electronic instruments contributes to developing new modalities of sound. The synthesizer does not substitute mechanical or electrical instruments because it is different by nature. The great quantity of different sounds results from the detuning of oscillators that causes its limited suitability for realizing tonal music. If a synthesizer or a sound is technically outdated, I arrange the piece of music it was performed with by choosing different sounds. I already consider this during the process of composing and producing. If I separate the composition from the sound, I am able to adjust it to changes in time and environment including the performance. Bionic research, the findings from the realm of psychoacoustics and electronic software and hardware contribute to the development of new electronic instruments. By analyzing the scientific properties of sound, the technique is improved and becomes more appropriate for composing tonal music and producing and combining precise frequencies. When electronic instruments perform the music of the spheres, they realize the tonal models and cosmological frequencies. The virtual studio technology is reduced to separate instruments that realize specific sounds and I alter multiple parameters that affect the physical level and the sound.

Mathematics and physics

The beauty of a composition and a mathematical procedure is based on logic, proportion and creativity. As a composer and mathematician, I think creatively and see a mathematical structure as esthetic and artistic. Is it reasonable to say that a fugue is correct from today's point of view? When natural scientists from the realms of mathematics, music and physics discover innovations, it influences the compositional technique of musicians. I evaluate references of music to other sciences from an artistic and sonic point of view. Although the art of music is related to multiple disciplines, it is unique and by composing I create an input that is based on mathematical and logical thinking. Because of the dual nature of music, the listener experiences the output cognitively and by sensory perception, and physical waves cause psychoacoustic effects. This relationship between mathematics, logic, sound engineering, physics, neuroscience and psychology is essential for the research of the music perception and neuroinformatics. The context between the mathematical structure of the composition and the psychoacoustic effects influences the research of artificial intelligence, biology and psychology. The acoustic shapes the electronic music and its equipment. The systematic and mathematical nature of music facilitates the integration of musical distributed artificial intelligence systems into the process of composing. The coder and the computer cooperate and contribute to the composition.

If I generate harmonies by applying or programming a musical distributed artificial intelligence system, I refine the procedure with an automatic mathematical procedure or one semi-automated by listening, evaluating and correcting. In both cases I predefine the esthetic concept, and because it is based on a consonant form of harmonics, I choose to include alterations in a more dissonant context which depends on the character of the piece of music. I define the threshold of the optimization function that is part of the compositional conception. Creativity is not a privilege of the arts and logic is not a privilege of mathematics but both are capacities and modes of thinking in general. Programming algorithms for computer music determines the compositional process. The application or coding of artificial intelligence software which defines the technical parameters and the process of composing music requires creativity. Classification algorithms support the design of sound and contribute to the configuration of hardware devices and the production of music. As a composer, I translate my conception, which is then made comprehensible by the computer and vice versa.

According to ancient Pythagorean and Egyptian science, arithmetic, music and astronomy are related, and music is a sibling of mathematics. Recurring patterns, numerical series and the periodicity of the cycles of the celestial bodies connect music and cosmology. As a musician, I translate the mathematical proportions and the harmony of the spheres into a compositional concept.

Until the twentieth century the tonal system was based on symmetries and scales that build a perfect system of frequencies and intervals with a particular proportion. The calculation of the frequencies is part of the realm of physics, music theory and composition. The compositional process consists of creating the tones and in defining a strategy to determine the frequencies in a serial form. The resulting frequencies coincide with the proportional measures that are calculated in a mathematical and cosmological way.

Musical theorists and developers of musical instruments of all ages create tonal systems and instruments that enable the composer to realize frequencies and sequences of sounds. Frequencies of light waves, sound waves and strings are calculated and transformed mathematically. I transfer the analogue or deviant behavior of the oscillations into the realm of music in a calculative way. The large number of harmonic oscillators of the string in quantum space time is a new model for the creation of proportions. The composition of music is part of the aural realm and the application of non-musical data and equations to the algorithm is artistic if it enables me to create sounds and forms of music that are sophisticated in comparison with the oeuvre of the masters of this craft.

Interdisciplinary creativity

Eai music integrates the sciences of physics, mathematics, music theory, composition, esthetics, audio engineering, electronic engineering and coding. These fields of knowledge are separate from one another and I apply my creativity to connect them. Interdisciplinary thinking requires the ability to connect areas that are distant from each other. When I combine two scientific realms, I develop an abstract conception of both of them separately, which requires a deep understanding of each of the structures, and I assimilate them in the form of their abstract version. On a technical level, I concretize the result. The rules of both fields of knowledge are applicable.

Algorithmic composition

The nature of music is a secret that scholars of different realms have not managed to uncover. Music is a science and an art form which develops dynamically and a definition of its complex nature does not exist. At the same time, there are multilayered rules that define the characteristics of good music and that scientists and artists develop in the course of music history. I create the sonic art of the future that encapsulates the past. This oeuvre belongs to the most sophisticated compositions of historical and contemporary music. The art and science of electronic algorithmic music designates the method of composing, audio engineering and programming and these procedures depend on each other. I combine the musical, esthetic and technical elements of a composition.

Music is the musician's language. Artificial intelligence procedures modify the ways musicians compose with algorithms. Programming algorithms and neural nets, controlling them by integrating physiological or scientific data or implementing electronically alienated noise is part of my compositions as far as I can ignore it, and the particular syntax, semantic content and magic of music is not affected. Algorithmic compositions relate the art of mathematics and music. The algorithm makes use of domain knowledge and of the rules that I design for the individual structure of my music.

Musical scientists apply quantitative research techniques and use a symbolic system that is connected with the physical nature of sound. By applying algorithmic methods, I realize the sound systems and sonic proportions of compositions. The goal of the technical procedure is the sonic realization of the beauty of music.

Algorithms are appropriate for both classical and quantum computers. Particular software is developed when music is composed with a quantum computer. The procedure is semi-automated and allows the composer, the software engineer and the computer to interact.

The model of algorithmic music is aural. It is the basis for the musical meta level, which is the result of the historical and contemporary development of music and its algorithms. The cosmological analysis of electromagnetic waves indicates a connection with the tonal system and its frequencies, and I integrate this relation into my compositions. These sounds of the planets are converted into audible frequencies. I blend the frequencies of brain waves and light waves which are non-aural models. They are physically similar to audio waves and I transform their periodicity mathematically. Mathematical models are based on numerical sequences and algorithms. Pitch-class theory contributes to the modeling of mathematical contours of sound sequences. I convert these models from other artistic and scientific realms into a language of music that is self-contained and consistent.

Floating counterpoint music is an algorithmic and processual compositional method integrating contrapuntal procedures. I compose stretti that consist of contrapuntal superimpositions of the soggetto to compress the texture and to increase the simultaneous complexity. I connect the main beats of the measure with the harmonies and the resolution of dissonances. Lengthy breaks do not function as caesuras because they do not bear a meaning. I structure my music by applying these principles. Symmetry rests on the microform and the macro-form. In addition to cadences and caesuras, the formal division structures the form of a piece of music. Elements which divide the musical space and time create patterns, and I apply multiple models. I compose forms that depend on numerical series, for example fractal series, and assimilate the model in the rules of music theory. I translate these procedures into musical language and its principles of harmony, counterpoint and soggetti that participate in the rhythm. The composition is a self-contained musical system. The listener recognizes the regularity, especially in the case of a symmetrical order and the repetition of parameters.

What does quantum music based on quantum logic sound like? The quantum computer solves particular tasks more efficiently than a classical computer and the composer can create particular quantum algorithms that are appropriate for the specific properties of the qubits. I use algorithms or graph algorithms to find the answers to logical and mathematical questions. The complexity of musical compositions can be NP-complete and it is unclear whether a quantum computer is able to perform them. As a composer, I restrict the procedure in order to use a tractable neural net which can be a quantum neural net. In this case, music is the result of training an artificial intelligence procedure. The machine learning technique generates content and applies statistical values that are the result of the analysis of a large number of pieces of music. The artificial intelligence procedure evaluates and classifies musical data of compositions that have already been created. For me as a composer, it is interesting to know which combinations of sounds and musical structures are familiar in the literature and which are exceptions. By applying artificial intelligence processes to analyze contemporary and historical compositions, I determine statistical values for the coordination of sounds and the arrangement of scales. This technique does not contribute to discovering which sounds or formations of sounds do not appear in the musical literature. I develop the musical structure of floating counterpoint by analyzing the probability of the sequence of tones and by evaluating and transforming the result.

I use the rules of music theory to structure the simultaneous or the successive sounds of frequencies. Although I relate nearly every sound of the austere form of floating counterpoint music to other sounds, the composition is at no point determined. I arrange tones or harmonies according to musical, mathematical and logical concepts that are based on axioms and inference rules. When I create a multi-agent system, I set rules that define the harmonic, formal, melodic and immanent rhythmic structure of the composition. From an artistic point of view, I select sounds and harmonies which are interpreted ambiguously. A technique for programming software that copes with the combined decision problems and complex ambiguities is currently unavailable and it is not possible to create sophisticated compositions in appropriate time or with suitable computational resources. When I interpret harmonies, I take into account their context and integrate them into my compositional conception. When I create plagal cadences to structure a piece of music that is not tonal in a functional harmonic sense, harmonies are more likely to be interpreted as referring to the subdominant sphere. The composer and the program alter single sounds if this is part of the conceptualization.

Based on my knowledge of musical history, I develop new rules for the processing of scales and forms, thereby creating a new genre. Capable computer programs participate in this procedure, which I apply to create new forms from inputs by applying rules to the arrangement of complex and ambiguous musical parameters. As an artist, I develop the composition by making choices regarding the musical parameters and especially the scales and contrapuntal processing. When I create algorithmic music, I define rules that limit the ability of the computer to improvise. By generating an algorithmic scheme which contributes to the composition of musical patterns, I transform the original input. Compositions of this form are individual and are the results of cooperation between the program and the composer. Even if the computer does not participate in the process, music is the outcome of applied algorithms. For me as a composer, it is as significant to intervene in the procedure that the algorithm controls as it is not to interfere and to accept the independent function that creates the structure of the piece of music. The algorithm is independent and regulates compositional patterns and sound. Contrapuntal principles shape the process and they constitute rules of the genre. Each autonomous output of an algorithm is the starting point of a new musicological development, and I evaluate it in the context of the artistic and scientific conception of the floating counterpoint genre.

I apply artificial intelligence techniques to the realm of digital sound processing in order to formalize the pitch and the time of audio files, to optimize the sound and to edit and correct it. I configure the hardware by applying the values that result from an automated analysis of the room properties. The significance of a musical composition depends on the proportionate symmetry, regularity and beauty that is the sonic result of algorithmic procedure and its contribution to the present and the future of music.

Binaural beats

The analysis of the psychoacoustic perception of music affects contemporary compositions. This realm is also significant for the creation of hardware which supports a sophisticated music experience. The brain has the ability to imagine sounds which do not appear physically. This auditory illusion is the basis, for example, of the invention of the MP3 file format. By composing, I create an aural imagination that is based on the expected reaction of the listener and on a psychoacoustic analysis. Scientists of the realms of physics, psychoacoustics, neuroscience and music theory shape the composition of binaural beats by analyzing the spectral separation of sounds and the temporal structure of hearing. I integrate brainwave music and binaural beats into the compositional process. The sound of binaural beats is a pure tone sine wave with a low frequency and a small difference between both partial sounds for both ears. Alpha waves, beta waves, theta waves and gamma waves are neural oscillations that resemble the electrical waves of the brain. I integrate these irregular frequencies into tonal music by interpreting them as dissonances and alterations of the pitch of a tone. I transform binaural beats into a musical parameter and as a composer and audio engineer I emphasize their particularities and associate them with the entire conception of the composition.

Tonal systems

Music theorists have developed different harmonic systems which provide the tonal material for composition during the course of centuries. The diatonic or well-tempered musical systems include the diatonic systems of ancient cultures, Kepler's harmonic aspectarian, the logarithmic tonal spiral and the tonal systems of Asia, Arabia and Africa. Diatonic systems exist in a well-tempered form. In the well-temperament, the just scale is substituted by a regular division of intervals, which facilitates the polyphonic composition and the modulation from one mode to another. The proportion is the main element of diatonic and well-tempered systems. The musical intervals are connected with celestial periods and harmonies. The meaning of one tone depends on its relation to the other tones. A hierarchy of proportions exists, and this is the result of calculating the orbits of the planets, their mathematical relations and the physical behavior of the sound waves which avoid prime numbers. Prime numbers do not participate directly in the tonal and physical system, but they play an important role in the fields of music and mathematics. Primes are essential factors that participate in the system of the nonprime numbers. They are also benchmarks in the tonal system and they are integrated into the system of music and its physical foundation in an indirect way. Consonant intervals constitute the main element of my compositions. Dissonances are important components of the musical texture because I integrate them as alterations that shape the sound of the music.

Cosmic periods, the orbits of the planets, atmospherics, colors and deoxyribonucleic acid or DNA resonate and their frequencies are partially identical to musical frequencies. Calculations of the periods of colors and atmospherics in the higher frequency ranges confirm the ancient diatonic system and the celestial sound system. Research into electromagnetic waves and brain waves refines the music of the spheres. I blend the brain waves with the tonal system of the music of the spheres that is related to a standard pitch A at 432 Hertz. The frequencies are not limited to a single value but a frequency range exists for the oscillations. I conclude that the references of brain waves to the musical frequencies that take the average values as an initial value coincide with the octaves of cosmic proportions.

Mathematical and physical calculations and astronomical research contribute to the creation of contemporary tonal systems. I integrate research results from the realms of the neuro sciences and biology into my musical conception. New scientific findings contribute to the compositional conception of the floating counterpoint genre and the current epoch. This procedure is connected with the construction of the electronic, electro-acoustic and mechanical instruments that realize the sound. The floating counterpoint conception shapes the music of the contemporary epoch in the twenty-first century by creating a new theoretical system that structures the musical elements of harmony, soggetto and mode. The algorithm shapes the composition and I create neural networks, genetic algorithms, serial techniques and contrapuntal procedures to shape contemporary and future music.

Mirror

My oeuvre is the mirror of my dynamic selves and of my mindful and conscious personality.

Magic

Magic is the coincidence of science and inspirational, logical and creative energy. It is the main source of carrying out research, composing, producing, coding and performing music.

Outlook

In my algorithmic electronic music of the twenty-first century, I integrate multiple scientific and artistic fields. The science of musicology describes genres in a philosophic and sociologic way. Music theory, audio engineering, artificial intelligence, and neuroscience are natural sciences. My conception of a sophisticated composition coordinates these different realms creatively.

My musical conception for the twenty-first century does not deny or alter tonal frequencies but it reinvents the function of musical parameters. This procedure is valid for electronic music, music created with concepts of artificial intelligence and for music that is arranged for conventional mechanical instruments.

As a composer who develops techniques for the artificial and creative intelligence of machines, I research historical and contemporary music in order to create a contemporary and future epochal conception of music, its genres and techniques. I blend the levels of art and engineering and combine creativity, logic, science and music. During the compositional process, the musical flow interacts with the computer and contributes to the development of coherent forms and sophisticated techniques. Floating counterpoint music is the result of a comprehensive conceptualization of technical and theoretical advances.

Index

Recent Publications

Knowledge Engineering. Artificial Intelligence and Legal Logic

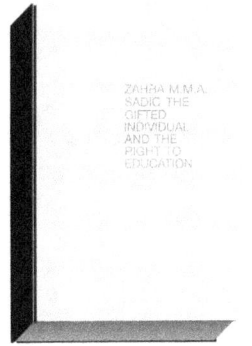

The Gifted Individual and the Right to Education

CDs

Anwar Ifriqia

Gardens of the Nile